AF270692

BABY ORANGUTANS

by Julie Murray

Cody Koala
An Imprint of Pop!
popbooksonline.com

Hello! My name is Cody Koala

This book is filled with videos, puzzles, games, and more! Scan the QR codes* while you read, or visit the website below to make this book pop.

popbooksonline.com/baby-orang

*Scanning QR codes requires a web-enabled smart device with a QR code reader app and a camera.

abdobooks.com

Published by Pop!, a division of ABDO, PO Box 398166, Minneapolis, Minnesota 55439. Copyright ©2024 by Abdo Consulting Group, Inc. International copyrights reserved in all countries. No part of this book may be reproduced in any form without written permission from the publisher. Cody Koala™ is a trademark and logo of Pop!.

Printed in the United States of America, North Mankato, Minnesota.

102023
012024

THIS BOOK CONTAINS RECYCLED MATERIALS

Cover Photo: Getty Images
Interior Photos: Shutterstock Images; Getty Images
Editors: Elizabeth Andrews and Grace Hansen
Series Designer: Candice Keimig

Library of Congress Control Number: 2023938793

Publisher's Cataloging-in-Publication Data
Names: Murray, Julie, author.
Title: Baby orangutans / by Julie Murray
Description: Minneapolis, Minnesota : Pop!, 2024 | Series: Baby animals | Includes online resources and index.
Identifiers: ISBN 9781098245252 (lib. bdg.) | ISBN 9781098245818 (ebook)
Subjects: LCSH: Animal babies--Juvenile literature. | Animals--Infancy--Juvenile literature. | Orangutans--Juvenile literature. | Apes--Juvenile literature. | Apes--Behavior--Juvenile literature.
Classification: DDC 591.39--dc23

Table of Contents

One Baby at a Time

Baby orangutans are born in nests high up in the trees. Mother orangutans often give birth to one baby at a time.

Watch a video here!

A baby orangutan weighs
four pounds (1.81kg) at birth.
It has **wispy** hair and a
wrinkled face. It drinks milk
from its mother.

Where Orangutans Live

There are Sumatran and Bornean orangutans. They live in the rainforests of each island.

The baby orangutan
has skinny arms and legs,
but its fingers are strong.

These help the baby **cling**
to its mother as she moves
through the treetops.

Chapter 2

Growing Up

A baby orangutan begins to eat soft fruit when it is three months old. The mother also chews up harder food and feeds it to her baby.

Orangutans eat fruit, nuts, leaves, and bugs.

Learn more here!

The baby begins to ride on its mother's back at two years old. As it grows, the mother teaches it how to find food and build a nest. At around four years old, the baby can climb and find food on its own.

A young orangutan will
continue to drink its mother's
milk until it is seven years
old. It usually stays with

its mother until the age
of ten. Females will visit
their mothers until they are
around 15 years old.

Female orangutans become adults around age 15. This is when they can give birth for the first time. Males **mature** around age 20. This is when they grow **throat pouches** and cheek pads.

Chapter 3

Up in the Trees

Orangutans eat, play, and sleep high up in trees. They move by climbing and swinging on branches. They build new nests every day to rest and sleep in.

Explore links here!

Using Tools

Orangutans are smart and can use tools. They use sticks to catch insects. They also use large leaves as umbrellas to stay dry.

Humans and orangutans share 97% of their **DNA!**

Complete an
activity here!

Making Connections

Text-to-Self

Imagine you were an orangutan. Would you like to live high up in the trees? Why or why not?

Text-to-Text

Have you read a book or watched a movie about other animals that live in a rainforest? What animals were in the book or movie?

Text-to-World

Young orangutans live with their mothers for many years. Can you think of other animals that live with their mothers for a long time?

Glossary

cling – to hold onto someone or something tightly.

DNA – short for deoxyribonucleic acid, a material in the body that helps determine what features a living thing will have and how they behave.

mature – to come to full or complete physical development.

throat pouch – located under the chin of male orangutans, a body part used to make a very notable and recognizable call that echoes through the forest.

wispy – resembling a wisp, or a thin bundle of hair.

wrinkled – having wrinkles or slight folds.

Index

Online Resources

popbooksonline.com

Thanks for reading this
Cody Koala book!

This book is filled with videos, puzzles, games, and more! Scan the QR codes* while you read, or visit the website below to make this book pop.

popbooksonline.com/baby-orang

*Scanning QR codes requires a web-enabled smart device with a QR code reader app and a camera.